GW00632686

AS Chemistry
UNIT 1

Salters

Module 2850: Chemistry for Life

Frank Harriss

To Maggi

Philip Allan Updates
Market Place
Deddington
Oxfordshire
OX15 0SE

tel: 01869 338652
fax: 01869 337590
e-mail: sales@philipallan.co.uk
www.philipallan.co.uk

This Guide has been written specifically to support students preparing for the OCR Salters AS Chemistry Unit 1 examination. The content has been neither approved nor endorsed by Salters or OCR and remains the sole responsibility of the author.

Typeset by Good Imprint, East Grinstead
Printed by Information Press, Eynsham, Oxford

Contents

Introduction

■ ■ ■

Content Guidance

■ ■ ■

Questions and Answers

Introduction

About this guide

This guide is designed to help you prepare for the first Salters AS Chemistry unit test, which examines the content of **Module 2850: Chemistry for Life**. This module is divided into two sections: **The elements of life** and **Developing fuels**.

The aim of this guide is to provide you with a clear understanding of the requirements of the unit and to advise you on how best to meet those requirements.

The book is divided into the following sections:
- This **Introduction**, which outlines revision and examination technique, showing you how to prepare for the unit test.
- **Content Guidance**, which provides a summary of all the 'chemical ideas' in Module 2850.
- **Questions and Answers**, in which you will find questions in the same style as in the unit test, followed by the answers of two students, one of whom is likely to get an A grade, the other a C/D grade. Examiner's comments follow these answers.

How to use this guide

- Read the section 'Revision and examination technique' in this Introduction.
- Decide on the amount of time you have available for chemistry revision.
- Allocate suitable amounts of time to:
  — each section of the Content Guidance, giving the most time to the areas that seem most unfamiliar
  — the questions in the Questions and Answers section
- Draw up a revision timetable, allocating the time for questions later in your timetable.
- When revising sections of the Content Guidance:
  — read the guidance and look at corresponding sections in your notes and textbooks
  — write your own revision notes
  — try questions from past unit tests and from other sources, such as *Chemical Ideas*.
- When using the Questions and Answers:
  — try to answer the question yourself
  — then look at the students' answers, together with your own, and try to work out the best answer
  — then look at the examiner's comments

Revision and examination technique

How do I find what to learn?

Well, we hope this book will be useful to you! Other sources are:

- the specification. This is the definitive one. If it's not in the specification it won't be in the paper! However, the specification is written in 'examiner-speak', so it might not always be absolutely clear what is required. This guide should help you to interpret the module content — every specification point is covered in the Content Guidance section.
- the 'Check your notes' activities in the *Activities* pack. These also suggest sources of details not found in the *Chemical Ideas* book. Some of the material is in the *Storylines* book and some in the activity sheets.
- your own and your teacher's notes. Preparation for an exam is not just something you do shortly before you take the paper. It should be an integral part of your daily work in chemistry. If you've left it a bit late this time, remember this when you are preparing for later units!

How much of the Storylines and Activities do I need to learn?

Have a look through for yourself, but you will find the details in the 'Check your notes' activity sheets referred to above.

The primary function of the *Storylines* book is to provide a framework and a justification for studying the theory topics. In the case of **The elements of life**, most of the theory is in *Chemical Ideas*. However, there is a lot more theory in the storyline of **Developing fuels**.

The activity sheets are provided to teach practical and other skills and to back up the theoretical ideas. However, they also contain some theory that does not occur elsewhere.

General revision tips

Revision is a very personal thing

What works for one person, does not necessarily work for another. You should by now have some idea about what methods suit you, but here are a few ways to set out your revision notes (to complement the 'Making the most of your study of chemistry' activity):

- mind maps — ideas radiate out from a central point and are linked together; some people like to colour these in
- notes with bullet points and headings
- small cards with a limited 'bite-size' amount of material on each

Make a plan

Divide up your material into sections (the Content Guidance section will be very helpful here). Then:

- work out how much time you have available before the exam

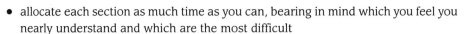

- allocate each section as much time as you can, bearing in mind which you feel you nearly understand and which are the most difficult
- fit this in with any revision your teacher is going to do — ask him or her for a summary

Write, write, write!

Whatever you do, make sure that your revision is *active*, not just flipping over the pages saying 'yeah, yeah, I know this already'. Write more revision notes, test yourself (or each other), *try questions*.

Test yourself

- Questions in *Chemical Ideas* are useful 'drill exercises' on topics, but are not all like exam questions.
- If you have taken end of unit tests, go through them again and then check your answers against the corrected version or the model answers you may have been given. These are much more like exam questions.
- Past papers are available and they give you a very good indication of what you will be facing.
- The Questions and Answers section of this book is designed for this purpose.

Know the enemy — the exam paper

I hope, since you will have prepared properly, you will be able to look on the exam as an opportunity to show what you can do, rather than as a battle! Be aware, however, that you must prepare yourself for an exam just as you would for an important sporting contest — be focused. Work hard right through the 75 minutes and do not dwell on difficulties — put them behind you. Try to emerge feeling worn out but happy that you have done your very best, even if you have found it difficult (others will probably feel the same way). Then forget it and don't have a post-mortem.

Every question tells a story

Salters is all about learning chemistry in relevant (i.e. real-life) contexts, so it is right that the exam questions should reflect this. Sometimes the context will come from *Storylines*, sometimes it will be a new one. Look carefully at the 'stem' (the introduction at the top of the question). Most of the important facts here will be needed somewhere in the question. Sometimes, small, additional stems are added later on. These are important too.

75 marks in 75 minutes

The evidence is that most people finish this unit test comfortably. However, if you do have trouble getting through papers (or if you tend to rush), plan to pace yourself through the paper so you can tell whether you are ahead of or behind the clock. There is a grid on the front of the paper that gives the marks for each question, which will be helpful here. It is best to work through the paper in order, from question 1 to question 4, since the first question is intended to be one of the easier ones.

50/25 knowledge/application of knowledge

This one you may not know about. Of the total, 50 marks test your knowledge and ask about things you will have learned. The other 25 marks are for the application of that knowledge to new situations or through doing calculations. These questions often begin with 'Suggest...' to make it clear that you are not expected to be able to recall the answer. There are about the same number of marks on the chemical ideas from **The elements of life** as there are on the chemical ideas from **Developing fuels**.

Easy and hard parts

The papers are designed so that, ideally, an A-grade candidate will get 80% (60 out of 75) and an E-grade candidate 40% (30 out of 75). The actual mark for each grade varies between papers, depending on the difficulty, and is only decided after all the papers have been marked. Some of the parts are designed with A-grade candidates in mind and so will seem quite demanding. Other parts are designed in order to allow an E-grade candidate to score 40% and so will seem rather easy. Thus, there are easy, middling and hard parts within each question.

Dealing with different types of question

Short-answer questions

These are the most straightforward, but remember:

- look at the marks available — make one good point per mark.
- look at the number of lines — this gives *some* idea of the length of answer required. Of course, handwriting differs greatly in size, but if you have written two words and there are three lines, you can assume you have not written enough to score full marks!
- don't 'hedge your bets' — if you give two alternative answers, you will not get the marks unless *both* are right. For example, if the answer is 'mass number' and you write 'mass number or relative atomic mass', you will score zero.
- read the question — don't answer a question that you have made up! Examiners do have kind hearts really, and they are genuinely sorry when they have to award zero for an answer containing good chemistry that is not relevant to the question asked. This is a problem with units that are examined twice a year. There are lots of past papers around, all asking slightly different questions on the same subject-matter. It's all too easy to give the answer to last year's question.

Long-answer questions

The same rules apply about marks, lines and reading the question. In addition:

- think before you write ('put brain into gear before operating hand') — perhaps jot a few points in the margin. Try to make your points logically.
- punch those points — if you have read any mark-schemes you will see that they give examiners advice on the weakest answer that will still just score the mark. Make sure your points are made well and win the mark without requiring a second's hesitation by the examiner.
- try to write clear sentences (though bullet points might be appropriate on some occasions).
- be sure you do not re-state the question, i.e. don't use words or phrases directly from the question as part of your explanation.

Command words in questions

A lot of care is taken in choosing which of these words to use, so note them carefully:

- 'state', 'write down', 'give' and 'name' require short answers only
- 'describe' requires an accurate account of the main points, but no explanation
- 'explain' requires chemical reasons for the statement given
- 'suggest' means that you are not expected to know the answer but you should be able to work it out from what you do know
- 'giving reason(s)' requires you to explain why you chose to answer as you did (if 'reasons' in the plural is stated, judge the number required from the number of marks)

Avoid vague answers

Sometimes it is clear that the candidate knows quite a lot about the topic but his or her answer is not focused. Avoid these words:

- 'it' (e.g. 'it is bigger') — give the name of the thing you are describing, otherwise it may not be very clear which object in the question is being referred to
- 'harmful' — if you mean 'toxic' or 'poisonous', say so!
- 'environmentally friendly' — say *why* it benefits the environment
- 'expensive' — always justify this word with a reason

Be careful with chemical particles — always think twice whenever you write 'particle', 'atom', 'molecule' or 'ion', and check that you are using the correct term.

If in doubt, write something

Try to avoid leaving any gaps. Have a go at every answer. If you are not sure, write something that seems to be sensible chemistry. As you will see from the Questions and Answers section, some questions have a variety of possible answers — the only answer that definitely scores zero is a blank.

Diagrams

You would be amazed at some of the diagrams examiners have to mark, so please:

- read the question. The answer is not always a reflux condenser! If it is an apparatus you know, then it is relatively straightforward. If you have to design something, look for clues in the question.
- make it clear and neat. Use a pencil and a ruler, and have a soft rubber handy to erase any errors.
- make sure it looks like real apparatus (which never has square corners, for example). Some apparatus drawn in exams would test the skill of the most proficient glass-blower.
- draw a cross-section, so that gases can have a clear path through. Don't carelessly leave any gaps where gases could leak out.
- think of safety. Don't suggest heating an enclosed apparatus, which would explode. If a poisonous gas is given off, show it being released in a fume cupboard.
- always label your diagram, especially if the question tells you to. Important things to label are substances and calibrated vessels (e.g. syringes or measuring cylinders).

Calculations

I'll let you into a secret — if you get the answer to a calculation right, the working does not need to be there (unless you could have guessed the answer). However, this is the most misleading piece of information in this book. It is always very easy to make mistakes, especially under the pressure of exams. So, set out the steps in your calculations clearly. Then you will get most of the marks if you make a slight mistake and the examiner can see what you are doing. Examiners operate a system called 'error carried forward' whereby, once an error has been made, the rest of the calculation scores marks if the method is correct from then on.

At the end of the calculation, there will be a line that reads, for example,

Answer_____   (2 marks)

Obviously you should write your answer clearly here! When you write down your numerical answer, check:

- **units** — most physical quantities have them (sometimes these appear on the answer line to help you)
- **sign** (remember oxidation states and ΔH values must be shown as '+' if they are positive)
- **significant figures** — you may be expected to analyse uncertainties more carefully in your practical work, but in exam papers all you have to do is to give the same number of significant figures as the data in the question

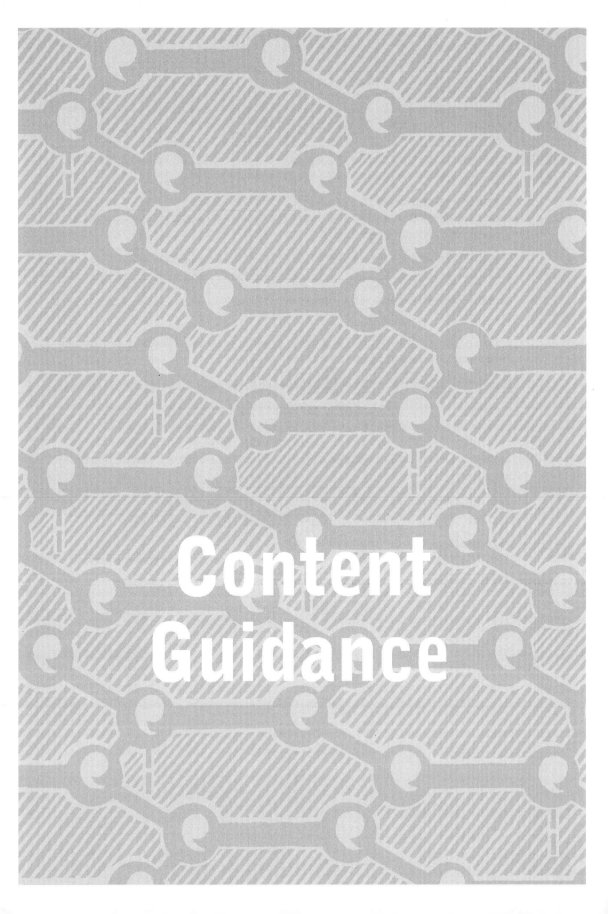

Content
Guidance

The material in this section summarises the chemical ideas from Module 2850: Chemistry for Life. It is arranged in a logical chemical order; not in the order in which you study it (which is determined by the content of *Storylines*).

Summary of content

About the atom: the make-up of an atom, nuclear reactions and radioactivity, electron arrangement and atomic spectra.

The periodic table: history, periodic trends of physical and chemical properties, ionisation enthalpies and electron structure, similarities and trends in a group.

Bonding and structure: ionic bonding, covalent molecules, shapes of covalent molecules, metallic bonding.

Alkanes: the alkane series, isomerism, naming alkanes, models, other organic compounds, naming alcohols.

Petrol: characteristics of a good fuel, improving alkanes, octane rating, pollutants.

Catalysis: heterogeneous catalysts in making fuels and in catalytic converters.

Entropy: what is entropy?

Mole calculations and equations: moles, formulae, equations, calculations from equations.

Enthalpy changes: measuring enthalpy changes, standard enthalpy changes, Hess's law, bond enthalpies, enthalpy changes of combustion in a homologous series.

How much of this do I need to learn?

The answer is, virtually all of it. It has been pared down to the absolute essentials. If you need any more detail on any aspect, you should look in your textbooks or notes.

About the atom

The make-up of an atom

The nucleus of an atom consists of **protons** and **neutrons**. Round the nucleus move the **electrons**. These are related as shown in the table below:

Particle	Mass	Charge
Proton	1	+1
Neutron	1	0
Electron	Very small	−1

The number of these particles in an atom is indicated like this:

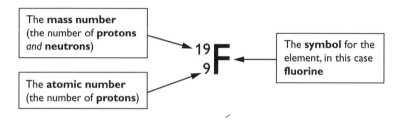

For a neutral atom, the number of electrons is equal to the number of protons.

Isotopes are forms of the same element. Their atoms have the *same number of protons* but *different numbers of neutrons*. For example, hydrogen has three isotopes: ^1_1H, ^2_1H and ^3_1H. These all have one proton (otherwise they wouldn't be hydrogen) but they have zero, one and two neutrons, respectively, as you can see from the mass numbers.

The **relative atomic mass** of an atom is defined as the number of times one atom is heavier than one-twelfth of an atom of carbon-12 (the isotope $^{12}_6C$). For an element that consists of just one isotope (for example, fluorine), this means that the relative atomic mass is virtually the same as the **mass number**: in this case, 19.

However, when an element consists of two or more isotopes, the relative atomic mass is the mean (average) of the **relative isotopic masses** of the isotopes, taking into account their **relative abundances** (how much of each of them occurs naturally). For example, chlorine has two isotopes, $^{35}_{17}Cl$ (which makes up 75% of chlorine) and $^{37}_{17}Cl$ (which makes up the remaining 25%). The relative atomic mass is calculated like this:

$$\frac{(35 \times 75) + (37 \times 25)}{100} = 35.5$$

Practise these calculations!

The masses of atoms can be measured using an instrument called a **mass spectrometer**. This works as shown in the following diagram.

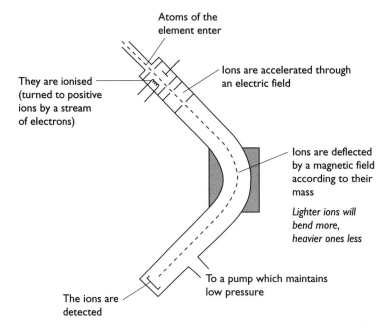

Tip You are more likely to be asked to *label* the diagram than to draw it.

The magnetic field is varied so that different masses hit the detector at different times. This enables a mass spectrum to be produced. The mass spectrum of bromine is shown below.

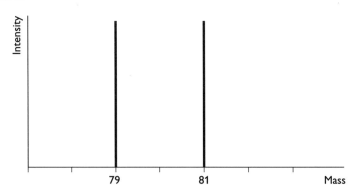

This shows that there are two isotopes, masses 79 and 81, of equal intensity. The relative atomic mass is given by:

$$\frac{(79 \times 50) + (81 \times 50)}{100} = 80$$

Nuclear reactions and radioactivity

The elements are made in stars like our Sun by **nuclear fusion**. This happens when lighter nuclei come together to make heavier ones. One example is

$$^1_1H + \, ^2_1H \longrightarrow \, ^3_2He$$

where ^1_1H etc. are called **nuclear symbols**.

This is an example of a nuclear reaction. The rule is: *the sum of the lower numbers (atomic numbers) must be equal on each side of the equation, as must the sum of the upper numbers (the mass numbers).* Any 'new' elements can be identified from the atomic numbers (for example, element number two is always helium).

Tip In the examination, you will always have the periodic table on your *Data Sheet* to consult.

Another reaction in which a heavier nucleus is formed is:

$$^9_4Be + \, ^4_2He \longrightarrow \, ^{12}_6C + \, ^1_0n$$

Here a neutron is also formed — notice how it is represented in such a nuclear equation. (It is very logical — no protons and one neutron!)

The protons and neutrons in some nuclei do not stay together. These nuclei are said to be **unstable** and **radioactive**. The properties of the three products of **radioactive decay** are shown in the following table.

Radiation[1]	What does it consist of?	Its penetrating power	Example of formation[2]
α-particle	Helium nucleus, ^4_2He	Small — stopped by paper and human skin	$^{235}_{92}U \longrightarrow \, ^4_2He + \, ^{231}_{90}Th$ Radioactive decay of uranium-235
β-particle	Electron, $^0_{-1}e$ (note that the electron comes from the nucleus — it is not one of the electrons that surround the atom)	Medium — stopped by sheets of metal such as aluminium	$^{14}_6C \longrightarrow \, ^{14}_7N + \, ^0_{-1}e$ Radioactive decay of carbon-14
γ-ray	Electromagnetic radiation	High — nothing really stops these rays but their intensity is usually reduced to safe levels by several centimetres of lead	Formed during α- or β-decay

[1] All three of these are ionising radiations. They carry a lot of energy so they can knock electrons off atoms in their path. This can cause damage to the genetic material in a cell, leading to cancer. Because they are ionising, they can be detected by a Geiger counter.

[2] The nuclear equations follow the rule about the sums of numbers given above. Notice the symbol for an electron: $^0_{-1}e$.

Tip You are not expected to learn the examples in the final column of the table.

Radioactive tracers

Radioactive tracers consist of radioactive atoms of an element which are used to follow the progress of the main bulk of the element, often around an animal body or plant structure. The position of the tracer can be found by the radioactivity it is emitting. Isotopes emitting β-particles, with their medium penetrating power, are often chosen for use as tracers. The half-life of the isotope used must not be too short, otherwise the radioactivity will not last long enough for detection. However, the radioactivity may be dangerous if it persists for long after the tracing has finished.

Electron structure and atomic spectra

The electrons in atoms are arranged in shells or energy levels. These determine the element's position in the periodic table, as shown below for the first 20 elements. The consequences of this will be considered further in the sections on 'The periodic table' and 'Bonding'.

H 1							He 2
Li 2.1	Be 2.2	B 2.3	C 2.4	N 2.5	O 2.6	F 2.7	Ne 2.8
Na 2.8.1	Mg 2.8.2	Al 2.8.3	Si 2.8.4	P 2.8.5	S 2.8.6	Cl 2.8.7	Ar 2.8.8
K 2.8.8.1	Ca 2.8.8.2						

Tip You will be expected to quote these electron structures but you will always have the periodic table in your *Data Sheet* for guidance.

The fact that electrons can exist only in definite energy levels leads to the production of atomic spectra. The energy levels of a hydrogen atom are shown below — other atoms' energy levels are similar, though more complex.

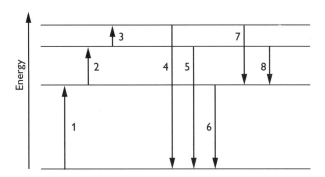

An electron in the lowest energy level (the 'ground state') can only go up to a higher energy level if it receives just the right amount of energy (1 in the diagram). Absorption of other specific amounts of energy moves the electron up further (2 and 3). If this energy comes from radiation such as light, it must have just the right frequency, the energy being proportional to the frequency.

Thus there are dark lines in the continuous ('rainbow-coloured') spectrum of the Sun because elements in the Sun's 'atmosphere' absorb certain frequencies. This is called an **absorption spectrum**.

After electrons have been excited (moved up energy levels) by being heated, they then drop down to lower energy levels again, giving out a frequency of light that is proportional to the energy difference through which they have fallen (e.g. lines 4 to 8). This gives rise to an **emission spectrum**, which consists of very fine coloured lines on a dark background. Part of the emission spectrum of hydrogen is shown schematically below (the numbers of the lines correspond to the energy changes in the energy levels diagram above).

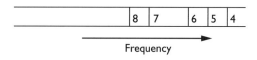

Frequency

Tip You may have to draw diagrams of energy levels and emission spectra like those above.

The periodic table

The periodic table lists the elements in order of **atomic number** and groups elements together according to their common properties.

History

In the late seventeenth century, chemists were trying to classify the elements. There were several attempts based on relative atomic mass but the most successful was that of Mendeleev in 1871. His table was very similar to that used today. However, he had to leave gaps for certain elements, which he said were not yet discovered, and he had to reverse the order of some elements (e.g. Te, I) so that they fitted the pattern of properties. Some people ridiculed him for this, but the discovery of Ga and Ge to fit two of his gaps began his vindication. Later, when atomic number was used to order the elements, this placed Te and I in Mendeleev's order naturally, so he was shown to be right here too.

Periodic trends of physical and chemical properties

Melting points and **boiling points** of the elements are said to show **periodic trends**, since the pattern of rise and fall is repeated as you go across the elements of each period. This is because the bonding of the elements changes from **metallic** to **giant covalent** to **covalent molecules** as one goes from left to right across a period (see the section on 'Bonding'). The **electrical conductivity** is high where there are metals on the left of a period but drops markedly on the right of a period where the elements are non-metals.

First ionisation enthalpies also show a periodic trend, as you can see in the chart below — this is explained in the next section.

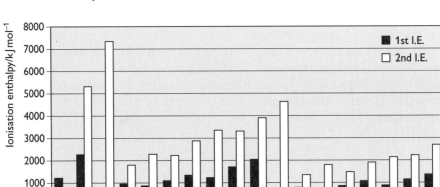

Tip You only need to know the outline *shapes* of these charts, *not* the values.

Ionisation enthalpies and electron structure

The **first ionisation enthalpy** measures the amount of energy needed to remove *one mole* of electrons from a mole of gaseous atoms to form the gaseous ion:

$$M(g) \longrightarrow M^+(g) + e^-$$

The **second ionisation enthalpy** measures the amount of energy needed to remove the *second* electron (*not* the first and the second) from *one mole* of singly-charged ions.

$$M^+(g) \longrightarrow M^{2+}(g) + e^-$$

The values of these depend on the **electron structure** (see page 16).

The ionisation enthalpy is *lower* (i.e. the ion is formed with greater ease) when:
- the electron being removed is further from the nucleus (smaller pull) — thus *the first ionisation enthalpy decreases down a group* as the outer electron shell gets further from the nucleus
- a smaller number of protons are attracting electrons in the same shell (smaller pull again) — thus *the first ionisation enthalpy increases across a period* as the number of protons in the nucleus increases

Similarities and trends in a group

Because elements in the same group have the same number of outer-shell electrons they tend to react in the same way. Thus they form ions with the same charge and molecules with the same general formulae. As we go down the group, the electrons are further from the nucleus and more easily lost. Thus the elements become *more metallic* (for example, the alkali metals) or *less non-metallic* (for example, the halogens).

Group 2

This group illustrates the above. As we go down the group from magnesium to calcium, strontium and barium, the elements become more metallic as their ionisation enthalpies become smaller. The areas you need to know about are:

- reactions with water
- the oxides and hydroxides
- the carbonates

Reactions with water

The elements get *more reactive* down the group:

$$M(s) + 2H_2O(l) \longrightarrow M(OH)_2(aq) + H_2(g)$$

The oxides and hydroxides

The oxides all react with water to form *alkaline* hydroxides:

$$MO(s) + H_2O(l) \longrightarrow M(OH)_2(aq)$$

The oxides and hydroxides are also *bases* since they react with acids to neutralise them. For example:

$$MO(s) + 2HCl(aq) \longrightarrow MCl_2(aq) + H_2O(l)$$

and

$$M(OH)_2(s) + 2HCl(aq) \longrightarrow MCl_2(aq) + 2H_2O(l)$$

The hydroxides become *more soluble* down the group.

The carbonates

All the group 2 carbonates decompose on heating:

$$MCO_3(s) \longrightarrow MO(s) + CO_2(g)$$

Barium carbonate needs the highest temperature, so it is said to have the highest **thermal stability**.

All the carbonates are insoluble but they become *less soluble* down the group (note this is the opposite way round to the hydroxides).

Bonding and structure

Ionic bonding

When a metal and a non-metal combine, they form an **ionic compound**. Elements combine in order to lose energy. Often the ions formed have the full electron arrangement of the nearest noble gas. Metals (being on the left of the periodic table) can most easily do this by *losing electrons* to form positive **cations**. Non-metals *gain electrons* to from negative **anions**. The charges on common ions are shown in the table below and are well worth learning. The charges on ions formed by a single element can be related to their position in the periodic table. Carbonate, nitrate and sulphate contain, of course, more than one non-metal and their formulae have to be learnt — but no chemist must be ignorant of these.

H H^+ and H^-							He
Li Li^+	Be	B	C	N N^{3-}	O O^{2-}	F F^-	Ne
Na Na^+	Mg Mg^{2+}	Al Al^{3+}	Si	P P^{3-}	S S^{2-}	Cl Cl^-	Ar
K K^+	Ca Ca^{2+}		CO_3^{2-} carbonate	NO_3^{2-} nitrate NH_4^+ ammonium	SO_4^{2-} sulphate		

When forming ionic compounds, atoms exchange electrons so that both gain a full outer shell of electrons. This can be shown by **dot–cross diagrams**, which show just the outer shell of electrons.

For example, magnesium has two outer-shell electrons. It can give one (shown as x in the diagram below) to each of two chlorine atoms to make magnesium chloride, $MgCl_2$.

$$\left[\begin{array}{c} \times\times \\ \times\, Mg\, \times \\ \times\times \end{array} \right]^{2+} \quad 2\left[\begin{array}{c} \bullet\bullet \\ \times\, Cl\, : \\ \bullet\bullet \end{array} \right]^{-}$$

Covalent molecules

When two non-metals combine they *share* electrons to form a **covalent** bond. Dot–cross diagrams can be used to illustrate this too. The vital things to remember are:
- the number of outer-shell electrons in an atom is the same as its group number
- the number of electrons in the outer shell of second-period atoms (Li to Ne) is limited to eight
- a complete outer shell of eight electrons is a very stable arrangement (hence the unreactivity of neon)

Tip You are only expected to know about compounds of the second period, or other compounds which resemble them.

Atoms share electrons to gain outer shells of eight (or two for hydrogen). Any non-bonding electrons that are not needed are shown as paired up (and are known as **lone pairs**). Examples are shown below. You should be able to draw these and others.

Methane   Ammonia   Water   Ammonium ion   Ethene

Note that in the ammonium ion, both electrons for one of the bonds (to the H^+ ion) come from the nitrogen. This is called a **dative covalent bond** (from the Latin 'to give', because both the electrons are *given* by the nitrogen atom). Ethene is a simple example of a molecule containing a **double bond**, which is shown as two shared pairs.

Shapes of covalent molecules

From the dot–cross diagrams, the shapes of molecules can be deduced by remembering the following simple rules:

- lone pairs, bonding pairs and double bonds all count as 'groups of electrons'
- groups of electrons repel each other and get as far away from each other as they can
- when there are *four* groups of electrons, the **bond angle** is approximately 109° (the angle of a regular tetrahedron). For example:

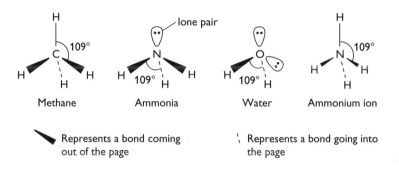

| Methane | Ammonia | Water | Ammonium ion |

◤ Represents a bond coming out of the page

╲ Represents a bond going into the page

- when there are *three* groups of electrons, the bond angle is approximately 120°. For example:

Boron trifluoride

Note: boron has only three electrons to pair up

Ethene

Note: the double bond counts as *one* group of electrons

- when there are *two* groups of electrons, the bond angle is 180°. For example:

Beryllium dichloride

Note: beryllium has only two electrons to pair up

Carbon dioxide

Metallic bonding

In the metal structure, the outer-shell electrons of metal atoms become detached from the atoms. Thus, metals consist of a **lattice** of positive ions. This accounts for their ability to bend and stretch without breaking, since the ions can move over each other in layers.

The electrons can move around between the ions and are said to be **delocalised**. This causes the metal to conduct electricity. Think of 'cation islands in a sea of electrons'. The positively charged ions attract the negative electrons and this attraction holds the structure together.

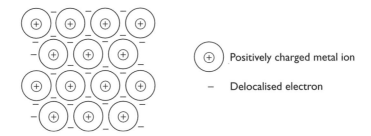

$\oplus$ Positively charged metal ion

$-$ Delocalised electron

Alkanes

The alkane series

Alkanes are the series of hydrocarbons with general formula $C_nH_{(2n+2)}$. This is an example of a **homologous series**. The first member of the series is **methane**. Three ways of representing the formula of methane are shown below.

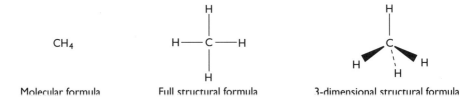

Molecular formula   Full structural formula   3-dimensional structural formula

Then come:

ethane, C_2H_6 heptane, C_7H_{16}
propane, C_3H_8 octane, C_8H_{18}
butane, C_4H_{10} nonane, C_9H_{20}
pentane, C_5H_{12} decane, $C_{10}H_{22}$
hexane, C_6H_{14} undecane, $C_{11}H_{24}$

Tip You should learn the names up to decane.

Isomerism

For butane and above, **structural isomerism** is possible. **Structural isomers** have the same molecular formula but a different structural formula. The structural, full structural and skeletal formulae of the isomers of butane and pentane are shown in the table below.

Name	Shortened structural formula	Full structural formula	Skeletal formula
Isomers of C_4H_{10}			
Butane	$CH_3-CH_2-CH_2-CH_3$		
Methyl-propane			
Isomers of C_5H_{12}			
Pentane	$CH_3-CH_2-CH_2-CH_2-CH_3$		
2-methyl-butane			
2,2-dimethyl-propane			

Notes on structural formulae

- **Shortened structural formulae** can be written in a variety of ways. They show the structure but they do not display all the bonds.
- **Full structural formulae** show *all* the bonds and atoms. Remember this if you are drawing one!
- **Skeletal formulae** look strange at first but they are used by chemists to represent larger structures. At the end of each line and at each angle there is assumed to be a carbon atom with the right number of hydrogens attached.

Naming alkanes

The rules for naming alkanes are:

- Identify the longest carbon chain. (This can sometimes 'go round corners'!) The number of atoms in this chain gives the basic name — propane, butane, pentane, etc.
- Use the names methyl (CH_3), ethyl (C_2H_5), etc., to describe the side-branches.
- Name the compound, giving the number of the carbon atom on which each side-branch hangs, counting from whichever end of the chain gives the smaller number. For example, 2-methylpentane means that a five-membered chain has a methyl group on the second atom along.
- When there are two branches on the same carbon atom, it is written like this: 2,2-dimethylpropane (as shown in the table on page 23).

Tip Note that numbers are separated by commas, numbers and words are separated by hyphens, and there is no gap between words.

Thus, the structure shown below is called 2,2,4-trimethylpentane.

Models

You will have used models in class and the examiners may want to test whether you understand these. One way they can do this is by giving you structures such as that shown in the diagram below and asking you to draw a full structural formula, as has been done.

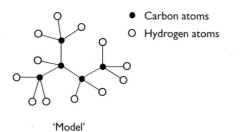

- ● Carbon atoms
- ○ Hydrogen atoms

'Model'

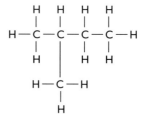

Full structural formula

Other organic compounds

As well as alkanes, you are expected to recognise the following homologous series.

Name	Shortened structural formula	Full structural formula	Skeletal formula[1]
Cycloalkanes (e.g. cyclohexane — rings of carbon atoms)			
Alkenes (e.g. propene — one double bond)	$CH_2{=}CH{-}CH_3$		
Alcohols (e.g. propan-1-ol — contain an $-OH$ group (see below))	$CH_3-CH_2-CH_2-OH$		
Ethers (e.g. 'diethyl ether' — contain a $C-O-C$ link)	$CH_3-CH_2-O-CH_2-CH_3$		

[1] In skeletal formulae, the bond to oxygen is shown and hydrogen atoms attached to the oxygen atoms are also shown. This is also the case for skeletal formulae in which there are nitrogen atoms in organic compounds.

Tip You are *not* expected to be able to name compounds apart from alkanes and alcohols (see below).

The compounds in the table above are all called **aliphatic** to distinguish them from compounds containing **benzene rings**, which are called **arenes** or **aromatic compounds**. Benzene has molecular formula C_6H_6 and is represented by the following symbol:

Naming alcohols

The rules for naming alcohols are:
- Find the longest carbon chain (as for alkanes). The number of carbon atoms gives the first part of the name: ethan, propan, butan, etc.
- The position of the OH group is described by the number of the carbon atom to which it is bonded, for example propan-2-ol, 3-methylbutan-1-ol, etc.

Petrol

Characteristics of a good fuel

Petrol consists of alkanes and other molecules. A good fuel:

- has the correct **volatility** to work in the engine
- has a high **energy density** (kJ kg^{-1}), i.e. it gives out a lot of energy per kilogram when it burns in air
- burns easily but does not 'auto-ignite' in the engine, i.e. has the correct **octane rating**
- does not corrode the engine
- produces the minimum level of polluting gases

Improving alkanes

Alkanes come from the fractional distillation of crude oil. This 'straight-run gasoline' can be made to perform better in engines by isomerisation, reforming and cracking.

Isomerisation means forming an isomer (usually more branched), for example:

$$CH_3-CH_2-CH_2-CH_2-CH_3 \longrightarrow$$

$$CH_3-CH-CH_2-CH_3$$
$$|$$
$$CH_3$$

Pentane
(boiling point 36°C;
octane rating 62)

2-methylbutane
(boiling point 28°C;
octane rating 93)

Reforming involves forming a ring compound from a chain (note that hydrogen is also produced), for example:

$$CH_3-CH_2-CH_2-CH_2-CH_2-CH_3 \longrightarrow$$

(cyclohexane ring structure) $+ H_2$

Hexane
(boiling point 69°C;
octane rating 25)

Cyclohexane
(boiling point 80°C;
octane rating 83)

Cracking means breaking up a long-chain alkane into shorter chains, some with double bonds and some branched, for example:

$$CH_3-CH_2-CH_2-CH_2-CH_2-CH_2-CH_2-CH_2-CH_2-CH_2-CH_2-CH_3$$
Dodecane (boiling point 196°C — far too high for petrol)

$$CH_3-CH-CH_2-CH_3 \quad + \quad CH_3-C-CH_2-CH=CH_2$$
$$|\qquad\qquad\qquad\qquad\qquad |$$
$$CH_3 \qquad\qquad\qquad\qquad\quad CH_3$$

2-methylbutane (boiling point
28°C; octane rating 93)

4,4-dimethylpent-1-ene (boiling point
63°C; octane rating 144)

Tip You need to know what is going on in these equations, but you are *not* expected to learn them.

All these processes require **catalysts** (see below). They produce compounds of a more suitable **volatility** and **octane rating**.

Octane rating

The **octane number** or **octane rating** of a petrol is a measure of how resistant it is to **pre-ignition** in a vehicle engine. In a modern, high-compression engine, the petrol is highly compressed in the cylinder before being sparked. Fuels of a low octane rating explode without being sparked (**auto-ignition**) and cause a juddering or knocking in the engine. From the above values, you can see that the octane rating of a fuel can be improved by:
- decreasing the chain length
- increasing the amount of chain branching
- making a ring compound
- making an unsaturated (double-bond) compound
- adding **oxygenates** (alcohols or ethers) which have high octane ratings

Pollutants

The following table lists and gives details of the main pollutants from vehicles.

Pollutant	Main effect	How is it produced?	How is it reduced?
Carbon monoxide, CO	Toxic (stops blood taking up oxygen)	Incomplete combustion of hydrocarbon fuels	Adjust fuel/oxygen ratio; use catalytic converters; adding oxygenates is also thought to help
Nitrogen monoxide NO (or NO_x)	Toxic; gives rise to acid rain, which damages plants and buildings	Nitrogen and oxygen from the air react in the heat of the engine	Use catalytic converters
Unburnt hydrocarbons	Combine with other gases to produce photochemical smog, which is a health hazard (especially for asthmatics)	Emission in exhaust gases and also from the engine and petrol tank; also released while filling the car	Adjust the fuel/oxygen ratio; better engine design; use catalytic converters; improved petrol tank and petrol pump design
Sulphur dioxide, SO_2 (or SO_x)	Gives rise to acid rain	Burning of sulphur compounds in fuel	Remove sulphur compounds from fuel (most are removed nowadays, so this is less of a problem)
Carbon dioxide, CO_2	Greenhouse gas	Burning of all hydrocarbon fuels	Use more energy-efficient fuels; use alternative non-hydrocarbon fuels

The use of hydrogen as a fuel would eliminate the emission of all of these, except for NO_x. The hydrogen, however, must be made, usually by electrolysis of sodium chloride solution. Production of the electricity for this in a conventional power station would itself cause pollution, particularly by sulphur dioxide. An alternative source of electricity (wind power, hydroelectric, etc.) is needed.

Catalysis

Catalysts and cars

A **catalyst** speeds up a reaction but is unchanged chemically at the end. Catalysts are used in two ways in connection with cars: in catalytic converters and in the manufacture of fuels.

- Catalytic converters, usually made of powdered platinum and rhodium on a ceramic support, are used to catalyse the oxidation of carbon monoxide and unburnt hydrocarbons to carbon dioxide (and water). They also catalyse the following important reaction which removes both carbon monoxide and oxides of nitrogen:

$$2NO + 2CO \longrightarrow N_2 + 2CO_2$$

- Catalysts are used to speed up cracking, isomerisation and reforming reactions.

In both the above cases, the reactants and products of the reactions are *gases* where the catalyst is a *solid*. Such catalysis (where the catalyst is in a *different state* from the reactants) is called **heterogeneous catalysis**. The mechanism of action of such a catalyst is often as follows:

- reactants are **adsorbed** (chemically bound) onto the catalyst surface
- bonds in the reactants weaken and break
- new bonds are formed between atoms from the reactant molecules
- the products diffuse away from the catalyst surface

A catalyst is said to be **poisoned** if another substance is more strongly adsorbed on to the catalyst surface and stops it functioning. Lead compounds would do this to catalytic converters and hence lead-free petrol had to be widely available before catalytic converters could be developed.

Entropy

What is entropy?

Entropy is a measure of disorder. The more ways of arrangement that are possible for a set of particles, the higher is its entropy.

A solid has least entropy as there are relatively few ways of arranging its particles (high order). Then, as we go from solid to liquid to gas, the entropy *increases*. Solutions often have high entropy too, as there are usually many different ways of arranging the particles.

Entropy increases in reactions in which more molecules of gas are produced.

A mixture of two substances has a higher entropy than the two substances separately. Again, there are more ways of arranging the particles of these substances when they are mixed.

Tip Try to name the particles if describing entropy, but *never* confuse molecules and ions.

Mole calculations and equations

Moles

A **mole** is a way of counting, like a 'dozen' or a 'score'. A dozen atoms would, of course, be 12 atoms. A mole of atoms is still a number but a much larger one: 600 000 000 000 000 000 000 000 atoms (much better written as 6×10^{23}). This large number is called the **Avogadro constant** (the number of particles in a mole).

Remember that the **relative atomic mass (A_r)** of an element is the number of times one atom of that element is heavier than one-twelfth of an atom of carbon-12. A_r for magnesium is 24. Thus, if there were 12 atoms of carbon in one beaker and 12 atoms of magnesium in another, the ratio of masses would be 1:2.

The mass of one mole of an atom is the A_r expressed in grams. So, for atoms,

$$\text{moles of atoms} = \frac{\text{mass (g)}}{A_r}$$

So, if enough carbon atoms were added to one beaker to make 12 g and the *same number* of magnesium atoms were added to the other beaker, their masses would still be in the ratio 1:2. There would now be 1 mole of each.

The **relative formula mass (M_r)** of a compound is the *sum of the relative atomic masses making up the formula.* For example, ethanol, C_2H_5OH, has

$$2 \times \text{C atoms} = 2 \times 12 = 24$$
$$6 \times \text{H atoms} = 6 \times 1 = 6$$
$$1 \times \text{O atom} = 1 \times 16 = 16$$
$$\text{Total} = 46$$

So, we say $M_r = 46$. This means that 46 g of ethanol contains the **Avogadro constant** of formula units, C_2H_5OH. (Since ethanol is made up of molecules, 46 can also be called the **relative molecular mass**, but **relative formula mass** is more general and applies to all formulae.) So, for compounds:

$$\text{moles of formula units} = \frac{\text{mass (g)}}{M_r}$$

Formulae

These can be worked out from combining masses by calculating the ratio of moles. Since all moles contain the same number of atoms, this gives the ratio of atoms.

Example 1

A hydrocarbon contains 3.0 g of carbon and 1.0 g of hydrogen. Calculate its formula. The formula can be worked out like this:

	C	H
Mass	3.0 g	1.0 g
Moles	3.0/12 = 0.25	1.0/1 = 1.0
Ratio (divide by smaller)	1	4

Thus the formula is CH_4. This is called the **empirical formula**, since it is the simplest ratio of the atoms. In this case, it is also the **molecular formula** of methane, that is, the actual formula of the molecule.

Example 2

Calculate the formula of a hydrocarbon containing 85.7% carbon. This can be done in a very similar way, provided it is realised that the remaining 14.3% must be hydrogen! The percentages mean that 85.7 g of carbon combine with 14.3 g of hydrogen.

	C	H
Mass	85.7 g	14.3 g
Moles	85.7/12 = 7.14	14.3/1 = 14.3
Ratio (divide by smaller)	1	2

Here the empirical formula is CH_2. Bonding theory tells us that this is not possible, so we try C_2H_4. This is, of course, possible, as are C_3H_6, C_4H_8, etc., so we see that CH_2 is the empirical formula of the alkene homologous series.

Example 3

What is the percentage of nitrogen by mass in the fertiliser ammonium nitrate (NH_4NO_3)? This is virtually the reverse of the previous method. First, work out the M_r of NH_4NO_3:

$$14 + 4 + 14 + 48 = 80$$

In this there are two nitrogen atoms (= 28). Thus,

$$\text{percentage of nitrogen} = \frac{28}{80} \times 100 = 35\%$$

Equations

These show that atoms are just re-arranged in chemical reactions, not created or destroyed. So, equations must *balance* — there must be the same number of each atom on each side. For example, methane burns in oxygen to form carbon dioxide and water. If you write the formulae of the reactants and products as

$$CH_4 + O_2 \longrightarrow CO_2 + H_2O$$

you can see the equation is *unbalanced*.

We cannot change the formulae of the individual molecules but we can change the number of moles reacting, that is, the big numbers in front of the individual molecules' formulae.

Putting '2' in front of H_2O will give four hydrogens on each side. The carbon atoms are already balanced, so now it is just the oxygens that are unbalanced. Having doubled the water molecules, there are now four oxygen atoms on the right, so we need $2O_2$ on the left to balance:

$$CH_4 + 2O_2 \longrightarrow CO_2 + 2H_2O$$

Sometimes, an odd number of oxygen atoms is needed in these circumstances. Then it is permissible to write the equation like this:

$$2C_3H_7OH + 9O_2 \longrightarrow 6CO_2 + 8H_2O$$

or

$$C_3H_7OH + 4.5O_2 \longrightarrow 3CO_2 + 4H_2O$$

Tip It *is* permissible to use halves (more so at AS than at GCSE). Note that when balancing the equation for the combustion of an alcohol, there is one oxygen atom in the alcohol formula. This is very easy to miss!

State symbols

The **state symbols** are:

- (g) gas
- (l) liquid
- (s) solid
- (aq) aqueous solution

State symbols add more information to an equation:

$$Mg(s) + 2HCl(aq) \longrightarrow MgCl_2(aq) + H_2(g)$$

Tip You need only add state symbols to an equation when the question specifically asks for them. Otherwise, it is safest to leave them out.

Calculations from equations

The point here is that the big numbers in front of formulae in the equations indicate how many moles are reacting.

So the rule is: *turn mass of the given substance into moles, use the ratio from the equation to calculate moles of the required substance, then turn the moles back to masses.*

Example 4

Calculate the mass of sodium carbonate which is formed when 16.8 g of sodium hydrogen carbonate is heated. (A_r: Na, 23; C, 12; H, 1.0; O, 16)

The equation is:

$$2NaHCO_3 \longrightarrow Na_2CO_3 + H_2O + CO_2$$

The M_r of $NaHCO_3$ is $(23 + 1 + 12 + 48) = 84$. So, moles of $NaHCO_3 = 16.8/84 = 0.20$.

The equation shows that two moles of $NaHCO_3$ form one mole of Na_2CO_3. So 0.20 moles will form 0.10 moles of Na_2CO_3. This has a mass of 10.6 g (M_r of $Na_2CO_3 = 106$).

A slight variation on this uses the fact that *one mole of molecules of a gas occupies 24 dm³ at room temperature and pressure* (you will be given this information when you need it).

So, in Example 4, if we needed to calculate the volume of carbon dioxide formed, we could have noted that the same number of moles of CO_2 are formed as Na_2CO_3, in this case 0.10. Thus, $0.10 \times 24 = 2.4$ dm³ of gas would be formed at room temperature and pressure.

If we are dealing with reactants and products that are *all gases*, the volumes are proportional to the moles reacting. Thus, for methane burning:

$$CH_4(g) + 2O_2(g) \longrightarrow CO_2(g) + 2H_2O(g)$$

Here, one volume of methane (say 1.0 dm³) would react with two volumes of oxygen (2.0 dm³) to form one volume (1.0 dm³) of carbon dioxide.

Enthalpy changes

When chemical reactions occur, heat may be given out or absorbed. The stored chemical energy that accounts for this is called **enthalpy**. **Enthalpy level diagrams**, such as those below, illustrate this.

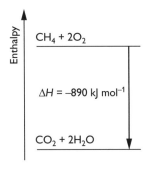

The combustion of methane
— **exothermic**

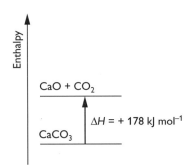

The decomposition of calcium carbonate
— **endothermic**

The *change* in enthalpy can be measured (directly for some reactions, or using **cycles** — see below). It is given the symbol ΔH.

- When enthalpy is *lost*, ΔH is *negative* and heat is *given out* — an **exothermic reaction**.
- When enthalpy is *gained*, ΔH is *positive* and heat is *taken in* — an **endothermic reaction**.

Measuring enthalpy changes

The enthalpy changes that occur during the combustion of liquid fuels can be measured directly using the apparatus illustrated below.

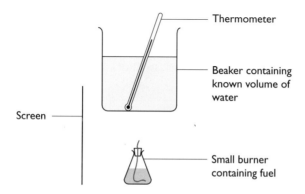

The temperature rise of a known mass of water is measured for a known mass of fuel burnt. The number of kilojoules given to the water can be worked out (mass of water $\times 4.2 \times$ temperature change) and divided by the number of moles of fuel to give the value in $kJ\,mol^{-1}$. $4.2\,J\,g^{-1}\,K^{-1}$ is called the **specific heating capacity** of water. It is a measure of the amount of energy which must be transferred to raise $1\,g$ of water through $1\,°C$ ($1\,K$). The results from this experiment are very inaccurate because of heat losses and evaporation of the fuel.

Standard enthalpy changes

Enthalpy changes vary with the conditions. **Standard conditions** are defined as:
- a *specified* temperature, usually $298\,K$
- one atmosphere pressure
- a concentration of $1\,mol\,dm^{-3}$

A *standard* enthalpy change (with the specified temperature shown) is written $\Delta H^{\ominus}_{298}$.

Standard enthalpy change of combustion, $\Delta H^{\ominus}_{c}$ is the enthalpy change that occurs when *one mole* of fuel is completely burnt in oxygen under *standard conditions*.

Standard enthalpy change of formation, $\Delta H^{\ominus}_{f}$ is the enthalpy change that occurs when *one mole* of a compound (in its *standard state*) is formed from its elements (in their *standard states*).

Hess's law

Hess's law states that *the enthalpy change in going from reactants to products is independent of the route taken*. This can be used to measure enthalpy changes of reaction *indirectly*. Two examples are shown below:

Example 1

$$C(s) + 2H_2(g) \xrightarrow{\Delta H_f} CH_4(g)$$

$$\Delta H_1 \searrow \qquad \swarrow \Delta H_2$$

$$CO_2(g) + 2H_2O(g)$$

Here ΔH_f is the enthalpy change of formation of CH_4, which cannot be measured directly. ΔH_1 is the sum of the enthalpy changes of combustion of one mole of carbon and two moles of hydrogen. ΔH_2 is the enthalpy change of combustion of one mole of methane.

By Hess's law:

$$\Delta H_f = \Delta H_1 - \Delta H_2$$

(Note: it is *minus* ΔH_2 since we are going *against* the arrow.)

Example 2

$$SO_2(g) + 2H_2S(g) \xrightarrow{\Delta H_r} 3S(s) + 2H_2O(l)$$

$$\Delta H_f\,(SO_2(g)) + 2 \times \Delta H_f\,(H_2S(g)) \nwarrow \qquad \nearrow 2 \times \Delta H_f\,(H_2O(l))$$

$$3S(s) + 2H_2(g) + O_2(g)$$

This is how the enthalpy change of a reaction can be measured using enthalpy changes of formation (which are available in tables).

$$\Delta H_r = -(\Delta H_f\,(SO_2(g)) + 2\Delta H_f\,(H_2S(g)) + 2\Delta H_f\,(H_2O(l))$$

Note:
- ΔH_f of an *element* is zero.
- It is vital to *multiply* the ΔH_f values of compounds by the number of moles in the equation.

Often a cycle is not asked for. Then the problem in Example 2 can be dealt with by a quicker method, using the equation:

ΔH_r = sum of ΔH_f (products) – sum of ΔH_f (reactants)

In Example 2, the ΔH_f values are as follows:

$$SO_2(g) + 2H_2S(g) \longrightarrow 3S(s) + 2H_2O(l)$$

| -297 | 2×-21 | 0 | 2×-286 | $kJ\ mol^{-1}$ |

So:

$$\Delta H_r = 2 \times -286 - (-297 + 2 \times -21) = -233\ kJ\ mol^{-1}$$

Tip There have been many questions requiring use of this equation. Practise these calculations.

Bond enthalpies

The **bond enthalpy** measures the energy required to break a bond into single atoms. It goes against intuition, but when bonds are formed, energy is *given out* so:

- bond-breaking is *endothermic*
- bond-making is *exothermic*

Thus *bond enthalpies* have *positive* signs.

The larger its bond enthalpy, the stronger the bond. Between the same atoms, double bonds are stronger and shorter than single bonds.

A Hess cycle can be drawn that shows how the enthalpy change of a reaction can be determined from bond enthalpies. For example, for the combustion of ethanol:

By Hess's law:

ΔH_r = bonds broken – bonds made

Tip There have been many questions requiring use of this equation. Practise these calculations.

- The number of C—C bonds broken is *one less* than the number of carbon atoms in the chain.
- Don't forget the oxygen double bond that is broken: if the equation has, say, 2.5 oxygen molecules, then multiply 2.5 by the bond enthalpy of O=O.
- It is easy to forget to double the numbers of moles when working out how many C=O and O–H bonds are made, for example wrongly using 2 and 3 (not 4 and 6) here.

Bond enthalpy calculations are very useful but they only give approximate answers because:

- they use *average* values of bond enthalpy for bonds in different compounds
- they rely on all reactants and products being *gaseous* (which is not always the case when they are in their standard states)

Enthalpy changes of combustion in a homologous series

The enthalpy changes of the straight-chain alkanes propane to dodecane are shown in the bar chart below.

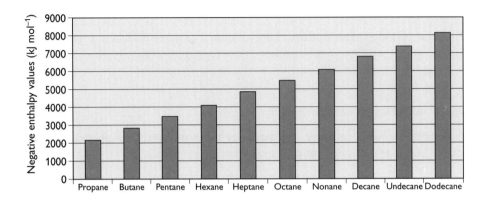

Tip You only need to know the *shape* of this, *not* the *detail*.

This shows that the differences between successive alkanes are very similar. This can be explained by the fact that they differ by CH_2. The combustion of an extra CH_2 group involves the same bonds being broken and made each time.

I n this section of the guide, there are four questions which between them test every lettered statement in the specification. They represent the kinds of question you will get in the unit test, in that they start with a context and they contain a wide range of subject matter from the whole specification. Unlike in the real thing, there are no lines or spaces left for the answers. Instead, the presence of a space or number of lines is indicated. The number of marks is, of course, also shown. However, taken together, these questions are much longer than a single paper, so do not try to do them all in 75 minutes.

After each question, you will find the answers of two candidates — Candidate A and Candidate B (using different candidates for each question). In each case, Candidate A is performing at the C/D level, while Candidate B is an A-grade candidate.

Examiner's comments
All candidate responses are followed by examiner's comments. These are preceded by the icon 𝑒 and indicate where credit is due. In the weaker answers, they also point out areas for improvement, specific problems and common errors.

How to use this section
- Do the question, giving yourself a time limit of a minute a mark; do not look at the candidates' answers or examiner's comments before you attempt the question yourself.
- Compare your answers with the candidates' answers and decide what the correct answer is; still do not look at the examiner's comments while doing this.
- Finally, look at the examiner's comments.

Completing this section will teach you a lot of chemistry and vastly improve your exam technique.

Cosmic rays and carbon dating

'Cosmic rays' from space consist of fast-moving electrons. These hit atoms in the upper atmosphere and cause them to emit high-energy neutrons. These neutrons react with nitrogen atoms to form protons and radioactive carbon-14 (which is used for 'radiocarbon dating').

(a) From the particles proton, neutron, electron, choose *two* in each case which:
    (i) have the same mass *(1 line)*   (1 mark)
    (ii) have a charge *(1 line)*   (1 mark)

(b) Use nuclear symbols to complete the equation for the reaction of a neutron with a nitrogen atom to form carbon-14 and a proton:
$$^{14}_{7}\text{N} + ^{1}_{0}\text{n} \longrightarrow \underline{\hspace{3cm}}$$
        (2 marks)

(c) The common isotope of carbon is carbon-12. Compare the nuclei $^{14}_{6}\text{C}$ and $^{12}_{6}\text{C}$ in terms of their nuclear particles.
    (i) State in detail what they both have in common. *(1 line)*   (1 mark)
    (ii) State in detail how they differ. *(1 line)*   (2 marks)

(d) (i) What name is given to the number 14 in $^{14}_{6}\text{C}$? *(1 line)*   (1 mark)
    (ii) What name is given to unstable isotopes such as $^{14}_{6}\text{C}$? *(1 line)*   (1 mark)

(e) The isotope $^{14}_{6}\text{C}$ decays with the loss of a β-particle. Write a nuclear equation for this process. *(space)*   (3 marks)

(f) Some other isotopes decay giving off an α-particle.
    (i) Give the nuclear symbol for an α-particle. *(1 line)*   (2 marks)
    (ii) Describe the difference in penetrating power between α-particles and β-particles. *(3 lines)*   (3 marks)

(g) While living creatures are alive, they exchange carbon with the carbon dioxide in the air. When a creature dies, this exchange stops and the carbon-14 decays. The half-life of the decay process (5370 years) enables the age of a once-living sample to be measured. The ratio of carbon-12 to carbon-14 in living things is 24 million to 1.
    (i) Suggest *one* way in which we exchange carbon with our surroundings. *(1 line)*   (1 mark)
    (ii) Suggest an instrument that could be used to measure the amount of carbon-14 in a sample by making use of the fact that it emits β-particles. *(1 line)*   (1 mark)
    (iii) Suggest a reason why the presence of carbon-14 is not used as a *tracer* to follow the path of carbon round the body. *(2 lines)*   (1 mark)

(h) Carbon also forms a stable isotope, carbon-13. The presence of this isotope can be detected by a mass spectrometer.
    (i) Label the diagram of the mass spectrometer below, writing your labels in the four boxes given. The labels should each describe what is happening to the particles at that point.   (4 marks)

1

question

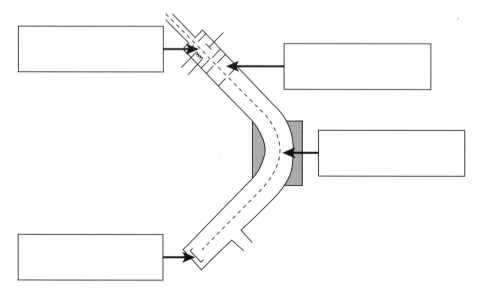

(ii) The dashed line shows the path of a carbon-12 particle through the machine. **Draw a solid line to show the path of a carbon-13 particle if the settings remained the same.** (2 marks)

(i) Naturally occurring carbon dioxide contains 1% carbon-13, the rest being carbon-12. **Calculate:**

  (i) The A_r of carbon in the sample to two decimal places. *(space)*
    Answer_____ (2 marks)

  (ii) The relative molecular mass of naturally occurring carbon dioxide to two decimal places (A_r: O, 16.00). *(space)* **Answer**_____ (1 mark)

(j) A long-term store of carbon on this planet is calcium carbonate ($CaCO_3$) in rocks. When this is heated, it gives off carbon dioxide and leaves calcium oxide.

  (i) **Write a balanced chemical equation** *with state symbols* **for the thermal decomposition of calcium carbonate.** *(space)* (2 marks)

  (ii) Calcium forms Ca^{2+} ions. **Explain how you could predict this from the position of calcium in the periodic table.** *(2 lines)* (2 marks)

  (iii) **Draw a dot–cross diagram (showing outer electron shells only) for the oxide ion O^{2-}.** *(space)* (2 marks)

  (iv) **Calculate the percentage by mass of carbon in calcium carbonate.** (A_r: O, 16.0; Ca, 40.0; C, 12.0). *(space)* **Answer**_____% (2 marks)

Total: 37 marks

■ ■ ■

Candidates' answers to Question 1

Candidate A

(a) (i) Protons and neutrons

  (ii) Protons only

Candidate B

(a) (i) Protons and neutrons

(ii) Protons and electrons

e Candidate B has answered both correctly. Candidate A scores 1 mark for part (i) but has become muddled in part (ii). The question would never be asked in this way if there were only one answer, so better to give two particles, even if you're not sure.

Candidate A

(b) $^{14}_{7}N + ^{1}_{0}n \longrightarrow ^{14}_{8}C + ^{1}_{1}p$

Candidate B

(b) $^{14}_{7}N + ^{1}_{0}n \longrightarrow ^{14}_{6}C + ^{1}_{1}p$

e Candidate B scores full marks. Candidate A has made a strange error in writing $^{14}_{8}C$, so he does not score the mark for this part of the equation, leaving just 1 mark.

Candidate A

(c) (i) Same number of protons

(ii) Different number of neutrons

Candidate B

(c) (i) Six protons

(ii) It has two more neutrons

e Candidate A is being vague! State *in detail* says the question, so he should have given the number of protons. No marks so far. In part (ii), there are 2 marks, so he will score 1 for saying they have different numbers of neutrons. Candidate B gives the correct detail in (i) but she makes the mistake of using 'it' in part (ii). To which isotope is she referring? Since it is impossible to tell from the question, she, too, only scores 1 mark.

Candidate A

(d) (i) Relative atomic mass

Candidate B

(d) (i) Mass number or nucleon number

e Candidate A is wrong. Relative atomic masses are not necessarily whole numbers. The top number in the nuclear symbol is the mass number, as stated by Candidate B. Has she, however, lost the mark by hedging her bets? Fortunately no, as these are alternative correct answers. Nevertheless, for such questions it is best to give just one answer, rather than two.

Candidate A

(d) (ii) Unstable

Candidate B

(d) (ii) Radioactive

e Candidate A's answer is not wrong but it gains no credit as the word 'unstable' is used in the question. Candidate B is correct.

Candidate A

(e) $^{14}_{6}C \longrightarrow {}^{14}_{5}B + {}^{0}_{-1}e$

Candidate B

(e) $^{14}_{6}C \longrightarrow {}^{14}_{7}N + e^{-}$

e Candidate A has made a fairly common error. Five minus one does not equal six on the bottom row! However, the symbol for the electron is correct and the atomic number 5 does correspond to boron, so he will score 2 out of 3 marks. Two marks also for Candidate B, who, for some strange reason, has not given the full nuclear symbol for the electron (Candidate A's is correct).

Candidate A

(f) (i) He

    (ii) α-particles do not penetrate far and β-particles penetrate further.

Candidate B

(f) (i) $^{4}_{2}He$

    (ii) α-particles will hardly penetrate through paper but β-particles penetrate thin sheets of metal.

e Candidate A has not written a *nuclear symbol* for the α-particle, so does not score for part (i). In part (ii), he has not given nearly enough detail. He should have seen that there were 3 marks available; also, the word 'describe' rather than 'state' should have given him a clue. He scores just 1 of the 3 marks here for realising that α-particles do not penetrate as far as β-particles. Candidate B scores the mark in part (i). She also scores 3 marks in part (ii) for giving details for both particles. 'Thin sheets of metal' is not an ideal answer, though; it would have been safer to add '...such as aluminium'.

Candidate A

(g) (i) Breathing out

    (ii) Mass spectrometer

    (iii) It is everywhere

Candidate B

(g) (i) Eating food

    (ii) Geiger counter

    (iii) Its half-life is so long that it would not give enough radiation.

e This is an 'application of knowledge' question, as the word 'suggest' shows. Candidate A has given a weak answer to part (i). A much better answer would be 'respiration', but Candidate A gets the mark, as this is not in the specification. (If in doubt, write something sensible.) Candidate B's answer is also correct. In part (ii), Candidate A's answer is ruled out by the mention of β-particles in the question. A mass spectrometer will not detect the tiny amounts of helium formed. A Geiger counter, however, will respond to each individual β-particle, so Candidate B gets the mark. There are lots of possible answers to part (iii), all of which require understanding of what a *tracer* means. Candidate A's answer is just acceptable as it implies that, since carbon-14 is in all life, it cannot be used to follow the passage of carbon. Candidate B's answer is better. Another possible answer is that the amounts of carbon-14 are too small. *Note that the 'stem' to part (g) gives a lot of clues here.*

Candidate A
(h) (i) and (ii)

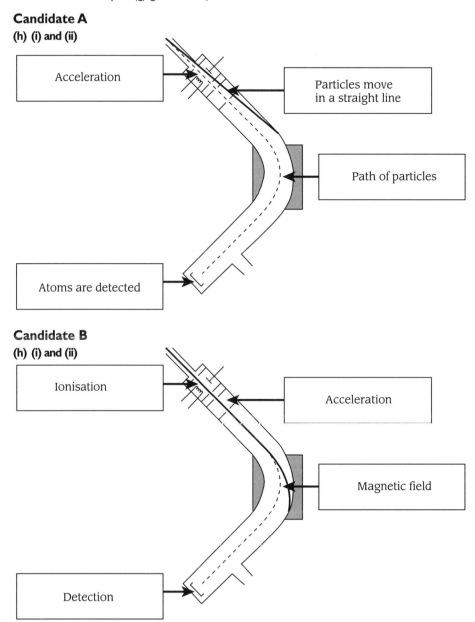

Acceleration

Particles move in a straight line

Path of particles

Atoms are detected

Candidate B
(h) (i) and (ii)

Ionisation

Acceleration

Magnetic field

Detection

e Candidate A has got muddled in part (i). The first three labels are wrong, and the last falls foul of the mark-scheme, which states that the mark cannot be awarded if the wrong kind of particle is mentioned. It should be 'ion' here. Candidate B is much nearer, though she loses a mark for labelling what is present in the middle ('magnetic field') rather than what is happening to the particles ('deflection'), so she gets **3 marks**. In part (ii), Candidate A realises that the heavier particle moves to the outside wall but not that this only happens under the influence of the magnetic field, so he does not score. Candidate B is correct, for **2 marks**.

Candidate A

(i) (i) 1% of 12 is 0.12. So the RAM is 12.12.

Candidate B

(i) (i) A_r C $= \dfrac{(99 \times 12) + (1 \times 13)}{100} = 12.01$ g

e Candidate A has gone off on the wrong track. He has tried a short-cut and failed. Candidate B has used the expression she has learnt and obtained the correct answer, 12.01. However, she has lost the second mark by adding the unit 'g'. Relative atomic masses have no units!

Candidate A

(i) (ii) $(2 \times 12.12) + 16.00 = 40.24$

Candidate B

(i) (ii) $12.01 + (2 \times 16) = 44.01$

e Candidate A has confused CO_2 with C_2O — no mark. Candidate B has done it right and scores **1 mark**.

Candidate A

(j) (i) $CaCO_3$ + heat $\longrightarrow$ CaO + CO_2

Candidate B

(j) (i) $CaCO_3(s) \longrightarrow CaO(s) + CO_2(g)$

e Candidate A should realise that 'heat' is not a chemical reagent and should not be written into equations. He would not score the first mark because of this and, of course, he would not score the second one as he has left out the state symbols. Candidate B has produced the correct answer, for **2 marks**.

Candidate A

(j) (ii) Calcium is in group 2, so it will form 2+ ions.

Candidate B

(j) (ii) Calcium has two electrons in its outer shell, which it loses to form Ca^{2+}.

e Each scores **1 mark** for part of the answer. Reference to both 'group 2' and 'two electrons in the outer shell' are needed for full marks.

Candidate A

(j) (iii) $\overset{\displaystyle ..}{\underset{\displaystyle ..}{:\,O\,:}}$

Candidate B

(j) (iii) 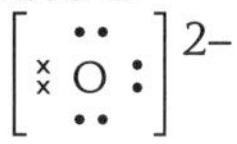 $\left[\begin{smallmatrix} \cdot\cdot \\ \times\ O\ : \\ \cdot\cdot \end{smallmatrix}\right]^{2-}$

💬 Candidate A has the idea that there are eight electrons around the oxygen and thus scores 1 mark. Candidate B's answer is completely correct and scores 2 marks. The inclusion of a pair of crosses, replacing one pair of dots, is not essential, but it indicates that two of the electrons were added to the oxygen atom to make the oxide ion.

Candidate A

(j) (iv) $CaCO_3 = 40 + (3 \times 28) = 124$

$\% = 12/124 = 9.7\%$

Candidate B

(j) (iv) $M_r\ CaCO_3 = 100$

$12/100 = 0.12 = 12\%$

💬 Candidate A has made a careless error again and assumed the formula is $Ca(CO)_3$. The rest of the working is right, though rather abbreviated, so he scores just 1 mark. Candidate B is correct, for full marks.

Hydrogen in the universe

The commonest element in the universe is hydrogen. The absorption spectrum of the gases surrounding the Sun shows that these gases contain hydrogen.

(a) An absorption spectrum consists of dark lines on a coloured background.
  (i) Explain what happens when hydrogen atoms absorb light. *(2 lines)* (2 marks)
  (ii) Explain why only certain frequencies are absorbed. *(2 lines)* (2 marks)
(b) All the other elements in the universe are made from hydrogen by nuclear reactions. What name is given to the *type* of nuclear reaction involved? *(1 line)* (1 mark)
(c) Hydrogen forms two ions, one of which is H^-. How many electrons are there in this ion? *(1 line)* (1 mark)
(d) In modern periodic tables, hydrogen is sometimes placed above group 1. The main reason for classifying hydrogen with group 1 is that it forms H^+ ions.
  (i) Why is hydrogen the first element in a modern periodic table? *(2 lines)* (2 marks)
  (ii) Suggest, with a reason, whether you would expect hydrogen to have been the first element in Mendeleev's periodic table. *(2 lines)* (2 marks)
  (iii) Write the equation, *with state symbols*, for the formation of H^+ which corresponds to the first ionisation enthalpy of hydrogen. *(space)* (2 marks)
  (iv) State, giving reasons, how the size of the first ionisation enthalpy of hydrogen would compare with the first ionisation enthalpy of lithium. *(2 lines)* (3 marks)
  (v) In view of your answer to part (iv), state, with a reason, which of lithium or hydrogen you would expect to form 1+ ions more easily. *(2 lines)* (1 mark)
  (vi) Unlike hydrogen, lithium and all the other group 1 metals show metallic bonding. Draw a labelled diagram which illustrates metallic bonding. *(space)* (3 marks)
(e) Sometimes, hydrogen is placed above group 7 in the periodic table. Suggest two similar ways in which hydrogen and fluorine form chemical bonds with other elements. *(3 lines)* (2 marks)
(f) Hydrogen forms a hydride with copper in which 4.45 g of copper combine to form 4.52 g of the hydride. Calculate the formula of the hydride. (A_r: Cu, 63.5; H, 1.0) *(space)* Answer_____ (3 marks)

Total: 24 marks

■ ■ ■

Candidates' answers to Question 2

Candidate A
(a) (i) Hydrogen atoms absorb light when they are excited to higher energy levels.

Candidate B

(a) (i) When electrons in the hydrogen atom absorb light energy, they rise to higher energy levels.

e Candidate A has left out a vital word — **electrons**! However, she has got the idea of moving up energy levels, and so scores 1 out of 2. Candidate B scores both marks.

Candidate A

(a) (ii) Only frequencies corresponding to the energy levels are absorbed.

Candidate B

(a) (ii) Electrons can only exist at certain energy levels. The energy differences between these levels require the absorption of definite frequencies of light.

e Candidate A is rather muddled and does not score. Note that it is the *differences* in energy level which require the absorption (or emission) of energy. Candidate B has once again produced a good answer, and scores 2 marks.

Candidate A

(b) Addition

Candidate B

(b) Fusion

e Candidate A is thinking along the right lines, but 'addition' is not used by chemists here, so no mark. **Fusion** is the correct term.

Candidate A

(c) 2

Candidate B

(c) None

e Here, Candidate A is right! There is one electron in a hydrogen atom, and another is added to make the 1– ion. Candidate B is presumably confusing the H^- ion with H^+, which, indeed, has no electrons.

Candidate A

(d) (i) Because it is the smallest atom

Candidate B

(d) (i) Because the periodic table is arranged in order of atomic number

e Candidate A's answer is far too vague! Candidate B scores just 1 — he should have added **and hydrogen has an atomic number of one/the smallest atomic number**.

Candidate A

(d) (ii) Not necessarily, because Mendeleev arranged the elements by their properties.

Candidate B

(d) (ii) Mendeleev arranged the atoms in order of atomic mass. Hydrogen is the lightest element.

Candidate A fails to score once again. Mendeleev arranged his periodic table by relative atomic mass and then noted that the elements in the columns had similar properties. Candidate B scores 2 marks, but it would have been safer to say **relative atomic mass** and **hydrogen is the element with smallest mass**.

Candidate A

(d) (iii) $0.5H_2(g) \longrightarrow H^+(g) + e^-$

Candidate B

(d) (iii) $H(g) \longrightarrow H^+(g) + e$

 Candidate A is wrong, in that ionisation enthalpies start from the **single gaseous atoms** of elements. However, she would get the state symbol mark. Candidate B is correct and scores 2 marks, though it is more usual to show the electron as **e⁻**.

Candidate A

(d) (iv) It would be lower as it has fewer protons to attract the electron.

Candidate B

(d) (iv) Hydrogen would have the higher ionisation enthalpy as the electron is closer to the nucleus.

 Candidate A has used the word 'it', which is dangerous! Does 'it' refer to hydrogen or lithium here? There are arguments for both! It doesn't matter in this case, as the answer is wrong anyway. Within a group, it is the distance of the electron from the nucleus which affects how tightly it is held. (Arguments about number of protons are used in discussing the change of ionisation enthalpy across *periods*.) Candidate B has named the element he is talking about and given a reason, for 2 marks. However, there are 3 marks available here and a more detailed reason is needed. He could have added '…**and is thus held more tightly**'.

Candidate A

(d) (v) Hydrogen, as it has the lower ionisation enthalpy.

Candidate B

(d) (v) Lithium, as its ionisation enthalpy is less than that of hydrogen. Thus, electrons are more easily removed.

 Candidate A has identified that it is the element with lower ionisation enthalpy which forms 1+ ions more easily. However, she has identified the wrong element and would not score the mark. Candidate B has given a full and correct answer (which actually gives more detail than is needed in this case — just the first sentence of his answer would do).

Candidate A

(d) (vi)

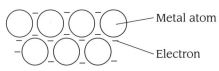

Candidate B

(d) (vi)

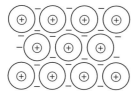

$\oplus$) positively charged metal ion

– Delocalised electron

The structure is held together by the attraction of the positive ions and the negative electrons.

e Candidate A has shown a regular lattice of metal particles and the electrons, so she scores 2 marks. She has not, however, indicated (or labelled) that the metal particles are ions. Candidate B has given a good answer. He has added the way in which the structure is held together, which is correct but not essential here. He scores all 3 marks.

Candidate A

(e) Hydrogen and fluorine form a single covalent bond together by sharing one pair of electrons.

Candidate B

(e) Either single covalent bonds or 1– ions.

e Candidate A has not read the question to its conclusion. It does not ask how hydrogen and fluorine combine with each other but how they combine with *other elements*. However, she scores 1 mark for talking about single covalent bonds. Candidate B has answered the question fully, for 2 marks.

Candidate A

(f) Moles of copper = 4.45/63.5 = 0.07
Moles of hydrogen = 4.52/1 = 4.52
The formula is CuH_{65}

Candidate B

(f) Moles Cu = 0.07
Mass H = moles H = 4.52 – 4.45 = 0.07
Formula is CuH

e Candidate A has failed to spot that the second mass is that of the *hydride*, not hydrogen. Still, she has worked out correctly the ratio of the moles she has calculated and scores 1 mark by 'error carried forward', though she ought to realise that this is a very unlikely formula! Candidate B has done the calculation correctly and scores full marks.

Isomers in petrol

Three hydrocarbons which are found in car petrol are pentane and its two structural isomers. These contribute different volatilities and octane ratings to the fuel.

(a) To which homologous series do these isomers belong? *(1 line)*   (1 mark)

(b) Draw the *full structural formula* of pentane (C_5H_{12}). *(space)*   (1 mark)

(c) Draw *skeletal* formulae for the *two isomers* of pentane and name them. *(space)*   (4 marks)

(d) (i) State how a fuel's *octane rating* affects how it performs in an engine. *(2 lines)*   (2 marks)

    (ii) Which one of pentane and its two isomers would have the highest octane rating? Give a reason for your answer. *(2 lines)*   (2 marks)

    (iii) Name a reaction of a hydrocarbon (apart from *isomerisation*) which will improve the octane rating. *(1 line)*   (1 mark)

(e) (i) Write a balanced chemical equation for the complete combustion of pentane. *(space)*   (2 marks)

    (ii) Use your equation and the bond enthalpy data below to calculate a value for the enthalpy change of combustion of pentane.

Bond	Bond enthalpy/kJ mol^{-1}
C–C	+347
C–H	+413
O=O	+498
C=O	+805
O–H	+464

    (space) Answer_____   (4 marks)

    (iii) The enthalpy changes of combustion of the two isomers of pentane are similar to your calculated value. Suggest why. *(3 lines)*   (2 marks)

    (iv) Which of the bonds in the table is the strongest? Give a reason for your answer. *(2 lines)*   (2 marks)

(f) An engine running on pentane would produce carbon monoxide and nitrogen monoxide. Describe how these pollutants arise in the engine and describe their polluting effects. *(8 lines)*   (6 marks)

(g) Carbon monoxide and nitrogen monoxide can be removed from the car exhaust by a catalytic converter containing a platinum/rhodium catalyst on a ceramic support.

    (i) What do you understand by the term *catalyst*? *(2 lines)*   (2 marks)

    (ii) Write a balanced chemical equation for the reaction of carbon monoxide with nitrogen monoxide to produce two less-harmful gases. *(space)*   (2 marks)

(iii) Explain in outline how carbon monoxide and nitrogen monoxide
react on the surface of the catalyst. *(space and 4 lines)* (3 marks)

(iv) Lead compounds in the petrol would poison the catalyst. Explain
what *catalyst poison* means. *(2 lines)* (2 marks)

Total: 36 marks

■ ■ ■

Candidates' answers to Question 3

Candidate A

(a) Hydrocarbons

Candidate B

(a) Alkanes

✏ The isomers are hydrocarbons, but then so is any substance containing just carbon
and hydrogen. This is too wide a statement to score. The correct answer is *alkanes*.
Learn to recognise the homologous series you need for this unit: alkanes, cycloalka-
nes, alkenes, arenes, alcohols and ethers.

Candidate A

(b)

$$H-\underset{\underset{H}{|}}{\overset{\overset{H}{|}}{C}}-\underset{\underset{H}{|}}{\overset{\overset{H}{|}}{C}}-\underset{\underset{H}{|}}{\overset{\overset{H}{|}}{C}}-\underset{\underset{H}{|}}{\overset{\overset{H}{|}}{C}}-\underset{\underset{H}{|}}{\overset{\overset{H}{|}}{C}}-H$$

Candidate B

(b) $H_3C-CH_2-CH_2-CH_2-CH_3$

✏ Candidate A is correct here. He has drawn a full structural formula showing all the
bonds and atoms. Candidate B has drawn a shortened structural formula, which
would not gain any credit. It is a correct representation of pentane but it *does not
answer the question!*

Candidate A

(c)

2-methyl butane   3-methylbutane

Candidate B

(c)

2-methylbutane   2,2-methylpropane

@ Candidate A has the right idea but he has made a lot of mistakes! First, there should not be dots on skeletal structures! Second, these two are the same structure. Third, there should not be a gap between methyl and butane in the name. He loses one of the skeletal formula marks for putting on dots (and he won't score the second one anyway). He scores 1 mark for the left-hand name, as the examiner would not penalise the gap. What about Candidate B? She has correctly identified the two structures and drawn correct skeletal formulae for them. However, she has made a mistake in the right-hand name. Although the '2,2' shows there are two groups on the second carbon, it is also necessary to add 'di'. Thus, the correct name is 2,2-*dimethylpropane*. She scores 3 out of 4.

Candidate A

(d) (i) The octane rating measures the engine's tendency to knock.

Candidate B

(d) (i) The octane rating measures how likely the fuel is to cause auto-ignition in the engine. The higher the rating, the less likely it is to do so.

@ Candidate A mentions knocking, which is permissible as an alternative to 'auto-ignition'. However, he fails to say which way round it works (the higher the rating, the *less* likely to knock), so he gains only 1 mark. Candidate B describes it in terms of auto-ignition and she also states the effect of a higher rating, for full marks.

Candidate A

(ii) The most volatile one

Candidate B

(ii) 2,2-dimethylpropane, because it is the most branched

@ Candidate A's answer is incorrect, and would have been too vague even if it had been the correct reason, so it does not score. Do not attempt to explain octane ratings by volatility (the ease with which a fuel evaporates). They are often related, the highest octane ratings going with the most volatile fuels, but a fuel does not have a high octane rating *because* it is volatile. Candidate B is correct once again and scores 2 marks.

Candidate A

(d) (iii) Adding an oxygenate

Candidate B

(d) (iii) Reforming

@ Candidate A has not read the question, which asks for a *reaction* of a hydrocarbon. Thus, his answer, while correct for a question such as 'how may the octane rating of fuels be improved?', scores nothing here. Candidate B was perhaps set on the right track by the word 'isomerisation' in the question. She had learnt this in the context of *reforming* and *cracking*, either of which would have scored the mark.

Candidate A

(e) (i) $C_5H_{12} + 11O_2 \longrightarrow 5CO_2 + 6H_2O$

Candidate B

(e) (i) $C_5H_{12} + 8O_2 \longrightarrow 5CO_2 + 6H_2O$

🖉 Candidate A scores 1 mark for getting the products — carbon dioxide and water — correct. However, he has made a balancing error, just adding the coefficients of CO_2 and H_2O to get the coefficient for oxygen, not realising that there is only one atom of oxygen per molecule of H_2O. Candidate B is right once again and scores 2 marks.

Candidate A

(e) (ii) Bonds broken

$5 \times$ C–C	1735
$12 \times$ C–H	4956
Total	6691

Bonds made

$5 \times$ C=O	4025
$6 \times$ O–H	2784
Total	6809

$\Delta H = 6691 - 6809 = -118$ kJ mol^{-1}

Candidate B

(e) (ii) Bonds broken

$4 \times$ C–C	4×347	1388
$12 \times$ C–H	12×413	4956
$8 \times$ O=O	8×498	3984
Total		10 328

Bonds made

$10 \times$ C=O	10×805	8050
$12 \times$ O–H	12×464	5568
Total		13 618

$\Delta H = 10 328 - 13 618 = -3290$ kJ mol^{-1}

🖉 Candidate A has made a good attempt here. Many weaker candidates do not give themselves a chance, just jotting down odd figures and arriving at an answer apparently at random! Because Candidate A has laid the working out quite well, it is possible to see the errors he has made and it is possible for the examiner to reward him when he has got things right! He has started by making a common mistake, thinking that there are five C–C bonds in pentane, when an inspection of his answer to part (b) would show him there are only four. He has also left out the O=O bonds. So, no marks so far! Under bonds made, he has made another common mistake in not noticing that there are *two* C=O bonds in a carbon dioxide molecule and two O–H bonds in a water molecule. Thus he loses the mark for bonds made. Then, however, he writes down the correct working to calculate ΔH (bond broken minus bonds made) and calculates it correctly. He should have been surprised at the small size of his answer, but he still scores the last 2 marks — 2 out of 4 — never to be sneezed at.

Candidate B is on a winning streak and gets all 4 marks. She has given even more detail (useful if you want to go back and double check) and got the calculation completely right.

Candidate A

(e) (iii) They are all really the same compound.

Candidate B

(e) (iii) They all have the same number and type of bonds.

e This is a difficult part and Candidate A has not grasped the idea. However, he has written something that might be relevant (always a good thing to do), though it does not score in this case. Candidate B has scored 1 of the 2 marks available. She is on track but does not give enough detail (signalled by the 2 marks and the 3 lines!). The second mark would be for saying that bond enthalpies do not vary much between compounds.

Candidate A

(e) (iv) C–H, since these bonds are the most unreactive.

Candidate B

(e) (iv) C=O, since it has the greatest bond enthalpy.

e Candidate A's answer is thoughtful, but incorrect. The reactivity of a bond is not a measure of its strength. Candidate B's answer is right, for 2 marks.

Candidate A

(f) When the fuel is not fully combusted, carbon monoxide is produced. This is a harmful gas, as is nitrogen monoxide which is formed in the engine when nitrogen compounds in petrol are not fully combusted.

Candidate B

(f) Carbon monoxide would be produced from incomplete combustion of the pentane. It is a toxic gas. Nitrogen monoxide is produced when nitrogen from the air reacts with oxygen in the heat of the engine. It is also toxic.

e Candidate A scores 1 mark for the point about the formation of carbon monoxide. He does not score for 'harmful'; **harmful to life** or **toxic** is much better. The formation of nitrogen monoxide explanation is incorrect, so Candidate A stops at 1 mark. Candidate B scores 1 for the formation of carbon monoxide. She also scores 1 mark for 'toxic', though she should have said that it stops the blood taking up oxygen. Her answer about the formation of nitrogen monoxide is also good, worth 2 marks. The question does not forbid stating the same polluting effects for the two compounds, but it would have been better to give different ones (e.g. nitrogen monoxide leads to the formation of acid rain, which damages trees and kills fish in lakes). Candidate B scores 5 out of 6, which is a pity as she probably knew enough to score full marks. She should have written a bit more and tried to give as much detail as possible.

Candidate A

(g) (i) A catalyst is a substance that speeds up a reaction by lowering its activation energy.

Candidate B

(g) (i) A catalyst speeds up a reaction but does not take part in it.

e Two rather different definitions! Who is correct? Well, Candidate A scores the 2 marks. Candidate B only scores 1 as she has confused 'chemically unchanged at the end' with 'not taking part'. You know that heterogeneous catalysts, for example, take part in the reaction (see below). So the mark-scheme will have wanted 'speeds up reactions' for the first point with either 'unchanged at the end' or 'lowers activation enthalpy/energy' for the second point. Let's suppose someone had written 'Speeds up a reaction by lowering the activation enthalpy, though it does not take part'. This would have scored 1 mark, as the incorrect last point would cancel out the correct second one. This may seem hard, but examiners do this to stop candidates giving a whole list of answers and hoping the examiner will pick the right one and ignore the rest. Bear this in mind — don't hedge your bets.

Candidate A

(g) (ii) $CO + NO \longrightarrow CO_2 + N$

Candidate B

(g) (ii) $2CO + 2NO \longrightarrow 2CO_2 + N_2$

e Candidate A scores 1 mark as he has identified the correct products (carbon dioxide and nitrogen) and written a balanced equation for their formation. However, he has made a bad error by forgetting that nitrogen gas in the air is N_2. Candidate B's equation is correct and scores 2 marks.

Candidate A

(g) (iii) The gases are absorbed on the surface where they react and the products leave.

Candidate B

(g) (iii) • The gases are adsorbed on to the surface of the catalyst.
  • Bonds are weakened and break.
  • New bonds are formed.
  • The products leave the surface.

e The questions says 'outline the stages', so an answer like Candidate B's is more likely to score, as it separates out the processes. Candidate A has missed the difference between *absorbed* (a physical process) and *adsorbed* (a chemical binding), so he loses the first mark. The next part of his answer is vague, though he picks up 1 mark for 'the products leave'. Candidate B scores all the marks, though she hasn't quite answered the question, which is about CO and NO, not gases in general. A perfect answer would have indicated *which* bonds were being made and broken. Note also that some space was left before the lines, so a diagram could have been used to illustrate these points.

question

Candidate A
(g) (iv) Lead blocks the active sight of the catalyst and stops it working.

Candidate B
(g) (iv) A poison is adsorbed onto the surface of the catalyst, so the reactant molecules cannot be adsorbed themselves.

Candidate A scores 1 mark for 'stops it working'. However, he needs a word more exact than 'blocks' (**adsorbed** or **binds**, for example), and he has confused catalysts with enzymes. A catalyst's surface is active all over, whereas an enzyme's has *active sites* (note the spelling!) where reactions occur. Candidate B starts well. She scores a mark for 'adsorbed onto the surface'. However, she has not made the point clearly enough that the catalyst stops working, so she does not score the second mark.

Uses of ethanol

Drinking alcohol can make you fat! This is because the body oxidises ethanol and uses much of the resultant energy to form fat. Ethanol can also be used in petrol, where it is called an oxygenate.

(a) (i) The enthalpy change of combustion of liquid ethanol can be measured by using burning ethanol to heat water. Draw a labelled diagram of a simple apparatus you could use for this. *(space)* (3 marks)

(ii) In such an experiment, the burning of 2.03 g of ethanol caused the temperature of 500 g of water to rise by 20.1°C. Calculate the energy transferred to the water, using the expression:

energy (in joules) = mass $\times$ 4.2 $\times$ temperature change

(space) Answer_____ J (2 marks)

(iii) Calculate the value this gives for the enthalpy change of combustion of ethanol in kJ mol^{-1}. (M_r for ethanol is 46.)

(space) Answer_____ kJ mol^{-1} (2 marks)

(iv) Suggest a reason (*apart from experimental inaccuracies*) why this method is unlikely to measure the *standard* enthalpy change of combustion of ethanol. *(3 lines)* (2 marks)

(b) The equation for the complete combustion of ethanol is:

$$C_2H_5OH(l) + 3O_2(g) \longrightarrow 2CO_2(g) + 3H_2O(l) \quad \text{(Equation 1)}$$

(i) You are given this equation and the standard enthalpy changes of formation of ethanol, carbon dioxide and water. Construct a Hess cycle which could be used to calculate the value of the standard enthalpy change of combustion of ethanol at 298 K. *(space)* (2 marks)

(ii) Use your cycle and the data given below to calculate the value of the standard enthalpy change of combustion of ethanol.

Compound	$\Delta H^{\ominus}_{f,298}$/kJ mol^{-1}
$C_2H_5OH(l)$	−277
$CO_2(g)$	−394
$H_2O(l)$	−286

(space) Answer _____ (3 marks)

(c) (i) Draw the *full structural formula* for butan-1-ol. *(space)* (2 marks)

(ii) The value of the standard enthalpy change of combustion of propan-1-ol is −2021 kJ mol^{-1}. Use this value and your answer to part (b) (ii) to estimate a value for the standard enthalpy change of combustion of butan-1-ol, explaining your method.

Estimate_____ kJ mol^{-1}

Explanation of method: *(3 lines)* (2 marks)

question

(d) Equation 1 also applies when ethanol burns in a car engine.

  (i) Calculate the volume of oxygen that would react exactly with $1.0\ dm^3$ of ethanol *vapour* according to Equation 1. (Assume both volumes are measured at the same temperature and pressure.) *(space)*

   Answer_____ (1 mark)

  (ii) Calculate the mass of water that would result from burning 2.3 g of ethanol. (M_r for ethanol, 46. A_r: H, 1.0; O, 16) *(space)* (3 marks)

(e) (i) Explain why liquid ethanol has a lower *entropy* than gaseous oxygen. *(2 lines)* (2 marks)

  (ii) In the reaction in Equation 1, would you expect the entropy to increase, decrease or stay the same? (Assume all the reactants and products are gases.) Give a reason for your answer. *(2 lines)* (2 marks)

(f) (i) Draw a dot–cross diagram for ethanol. *(space)* (2 marks)

  (ii) State the values of the following bond angles in the ethanol molecule:

   (1) $H\overset{C}{\frown}H$ (2) $C\overset{O}{\frown}H$ (2 marks)

  (iii) Explain your answer to part (f) (ii) (2). *(3 lines)* (3 marks)

(g) Suggest a reason why oxygenates such as ethanol are added to petrol. *(2 lines)* (1 mark)

(h) State, with a reason, whether you would class ethanol as an *aliphatic* or an *aromatic* compound. *(2 lines)* (1 mark)

Total: 35 marks

■ ■ ■

Candidates' answers to Question 4

Candidate A

(a) (i)

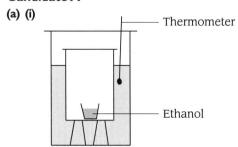

Candidate B

(a) (i)

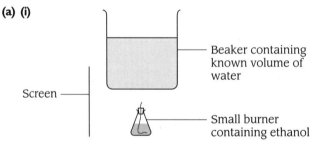

e Neither scores full marks. Candidate A has been confused by the memory of a bomb calorimeter. This apparatus would not work well, but Candidate A scores 1 mark for the thermometer in the water. This is the mark that Candidate B loses! He has, for some reason, left this out. The rest of the diagram is fine, so it scores 2 out of 3 marks.

Candidate A

(a) (ii) Energy = $2.03 \times 4.2 \times 20.1 = 0.17$ **J**

Candidate B

(a) (ii) Energy given out = $500 \times 4.2 \times 20.1 = 42\,210$ **J**

e Candidate A has not realised that it is the water which is being heated; thus the 'mass' referred to is the mass of the water, not the mass of the fuel. This makes her answer wrong from the start. She could, however, have started to score here by the principle of 'error carried forward', which means that the examiner will do the rest of the calculations with the candidate's wrong values and award the marks if the candidate's later answers agree with these. Here, however, Candidate A has also converted to kilojoules, which means she loses the second mark (for the answer), as the answer is required in joules (the 'J' is given at the end of the answer line). Candidate B has got it right. However, he has not checked on the significant figures. The question has data to three significant figures, so he should have given his answer as 42 200 (the 'J' is given on the answer line). He does not lose credit here, as his answer is carried through to part (iii), but it is safer to give all answers to the correct number of significant figures.

Candidate A

(a) (iii) $0.084 \times 46/2.03 = 1.9$ **kJ mol^{-1}**

Candidate B

(a) (iii) $\dfrac{42\,210}{1000} \times \dfrac{46}{2.03} = 956.25$ **kJ mol^{-1}**

e Candidate A has recovered her stride and scores full marks on the 'error carried forward' principle. Candidate B has used the right method. He has converted J to kJ by dividing by 1000 and then scaled up from 2.03 g to 1 mole (46 g) by multiplying by 46/2.03. He has, however, still given too many significant figures, so he loses the significant figure mark at this stage. The best answer is 956 kJ mol^{-1}.

Candidate A

(a) (iv) Heat is lost in the apparatus.

Candidate B

(a) (iv) Most of the water produced is vapour. The standard state for water is liquid.

e Candidate A has not read the question carefully. Heat loss is an experimental inaccuracy, so this answer does not score. Candidate B is correct *and* has given enough detail to score both marks.

Candidate A

(b) (i) $C_2H_5OH(l) + 3O_2(g) \longrightarrow 2CO_2(g) + 3H_2O(l)$

2C + 7O + 6H

Candidate B

(b) (i) $C_2H_5OH(l) + 3O_2(g) \longrightarrow 2CO_2(g) + 3H_2O(l)$

$2C(s) + 3O_2(g) + 3H_2(g)$

Candidate A has the right idea of using the elements but she has made several errors. First, the enthalpy changes of formation required start from the elements *in their standard states*, not single atoms as shown. Also, there are no state symbols. So Candidate A is awarded 1 mark for the right 'shape' of the cycle. (The fact that the left-hand arrow is in the wrong direction will be considered in the mark-scheme for the next part.) Candidate B's answer is correct, for 2 marks.

Candidate A

(b) (ii) Answer = 394 + 286 − 277 = 403 kJ mol⁻¹

Candidate B

(b) (ii) $\Delta H^{\ominus}_{C,298} = (2 \times -394) + (3 \times -286) - (-277) = -1369$ kJ mol⁻¹

Candidate A has again made several mistakes. First, she has failed to multiply the $\Delta H^{\ominus}_{f,298}$ values for CO_2 and H_2O by 2 and 3, respectively. Second, she has the signs round the wrong way. Even so, she has calculated the answer correctly from her values and has given the correct units. Does she get the third mark, therefore, with error carried forward? Unfortunately not, as she has failed to put in the plus sign. *Remember, ΔH values must have a sign, either plus or minus.* Of course, she should have realised that all enthalpy changes of combustion are negative and gone back and checked her work! Candidate B has the correct answer. He has multiplied the enthalpy changes of formation of CO_2 and H_2O by their respective numbers of moles and added them. He has then subtracted the enthalpy change of formation of ethanol, keeping a careful eye on the signs.

Candidate A

(c) (i) $CH_3CH_2CH_2CH_2OH$

Candidate B

(c) (i)

$$\begin{array}{c} \quad\;\; H \;\;\; H \;\;\; H \;\;\; H \\ \quad\;\; | \;\;\;\;\; | \;\;\;\;\; | \;\;\;\;\; | \\ H - C - C - C - C - O - H \\ \quad\;\; | \;\;\;\;\; | \;\;\;\;\; | \;\;\;\;\; | \\ \quad\;\; H \;\;\; H \;\;\; H \;\;\; H \end{array}$$

🖉 Candidate A has given a structural formula but not the *full* structural formula, which shows all the bonds and atoms. She scores **1** mark for showing the shape of the molecule. Candidate B scores **2**, 1 for the 'backbone' and 1 for the O–H group.

Candidate A

(c) (ii) Estimate: −4445 **kJ mol⁻¹**

Explanation: it follows the trend

Candidate B

(c) (ii) Estimate: −2672 **kJ mol⁻¹**

Explanation: the difference between the values for ethanol and propanol is −652, so the value for butanol is −2021 − 652 = −2673

🖉 There is a considerable difference between the answers: can both estimates be right? In fact they are, with error carried forward in the case of Candidate A. (*Moral: keep going, even if you think you've gone wrong earlier.*) Candidate B has got it completely right and explained the method, so he gets both marks. (Note that, had it said something like 'explain the theory underlying your method', it would have been necessary to talk about extra CH_2 groups, etc.) Candidate A, after a good start, has been far too vague. Clearly, she knew what to do, as she got the answer right, though she failed to communicate her method, leaving her just **1** mark.

Candidate A

(d) (i) $3 \times 24 = 72$ dm³

Candidate B

(d) (i) 3.0 dm³

🖉 Candidate A became confused and worked out the volume of three moles of gas — not what is required here. (Note, the 24 dm³ would always be given if you needed to use it.) Candidate B has seen that there is a 3:1 ratio between moles of oxygen and moles of ethanol vapour. Since all gas molecules occupy approximately the same volume (at the same temperature and pressure), all that is required is to multiply 1 by 3!

Candidate A

(d) (ii) $2.3/46 \times 18 = 0.9$ g

Candidate B

(d) (ii) Moles ethanol = 2.3/46 = 0.05

Moles water = 0.15

Mass water = $0.15 \times 18 = 2.7$ g

🖉 Candidate A has not made it very clear what she is doing and has made a mistake, so should she score any marks? Here the mark-scheme tells examiners to look for '2.3/46' and '× 18' to reward in the right contexts, so she scores **2** marks. Her only mistake is to forget that one mole of ethanol burns to form three moles of water. You might think this is generous and, I hope, learn to put in more working! Candidate B has given a good answer, for **3** marks, though he has not actually shown the working for the multiplication by three. *Give plenty of detail!*

Candidate A

(e) (i) There are more ways of arranging the molecules of a gas than there are of a liquid.

Candidate B

(e) (i) Entropy is disorder. Liquid ethanol has less disorder than gaseous oxygen.

e For the pleasure of those who like backing the underdog, Candidate A has given the better answer! The question asks for an explanation and she has given it. It would have been safer to have mentioned ethanol and oxygen, the specific examples, but there is enough here to score both marks. The crucial points are 'ways of arrangement' and 'molecules'. Candidate B's answer is weaker. The first sentence is an alternative way of scoring 1 mark but the second virtually re-states the question and does not add any new information, such as 'there are more ways of arranging the particles in a gas than in a liquid'.

Candidate A

(e) (ii) Increase, as there are more molecules.

Candidate B

(e) (ii) There are more molecules on the right-hand side of the equation, so the entropy will increase.

e Here Candidate A has been vague again and only scores 1 mark! More molecules where? Candidate B has given a full and correct answer, for 2 marks.

Candidate A

(f) (i)
```
       H  H
       x• x•
   H x C x C x O x H
       x• x•
       H  H
```

Candidate B

(f) (i)
```
       H  H
       x• x• ••
   H x C x C x O x H
       x• x• ••
       H  H
```

e Candidate A has nearly got it right and scores 1 of the 2 marks for the bonds between the atoms. However, she has forgotten that oxygen has six electrons in its outer shell (as it is in group 6) and so it has two *lone pairs* which must be shown. Candidate B has drawn the correct diagram, for full marks.

Candidate A

(f) (ii) (1) 109
      (2) 180

Candidate B

(f) (ii) (1) 109°
      (2) 109°

e Candidate A's answer to part (1) is right, if we excuse her for not putting the degree sign (as was done in a recent mark-scheme) — be careful, though! Part (2) is wrong and possibly follows from the omission of the lone pairs in the dot–cross diagram. Candidate A is awarded just 1 mark. Candidate B has them both right, for 2 marks.

Candidate A

(f) (iii) The groups of electrons repel each other.

Candidate B

(f) (iii) There are four groups of electrons (two bonding pairs and two lone pairs) which repel each other and maximise the repulsion.

e Candidate A is vague in not specifying the number of groups of electrons, but right to state that they repel, so she scores 1 mark. However, once again, she has given an imprecise answer. If she had continued to say '…and get as far away from each other as possible', she would have scored another mark. Candidate B is well on track…but, wait a minute, *maximise repulsion*? No! They get as far apart as possible to *minimise* repulsion. He therefore fails to score the third mark.

Candidate A

(g) To make them burn more cleanly.

Candidate B

(g) To increase the octane rating.

e Here are two very different answers. Can they both be right? In fact they are, and both score the mark, though it would have been safer for Candidate A to add '…producing less unburnt petrol *or* less carbon monoxide'.

Candidate A

(h) Aliphatic, because it has no double bonds.

Candidate B

(h) Aliphatic, because it does not have a ring.

e Both are correct to state that it is aliphatic, but neither scores any marks! Examiners cannot award marks for a 50/50 choice, so the 1 mark is for the reason, once the correct answer has been chosen. There are many unsaturated compounds that are aliphatic and there are some cyclic (ring) compounds (like cyclohexane) that are not aromatic. Aromatic compounds contain a *benzene ring*; aliphatic compounds do not.